Multimeter Mastery Handbook

A Comprehensive Guide to Electrical Measurement for Beginners

Peter J. MacCollin

Multimeter Mastery Handbook

Copyright © 2024 by Peter J. MacCollin

All Rights Reserved.

No part of this book may be used or reproduced by any means, graphic, electronic, or mechanical, including photocopying, recording, taping, or by any information storage retrieval system without the written permission of the publisher.

Multimeter Mastery Handbook

TABLE OF CONTENT

Introduction to Multimeters..3
Understanding the Basics... 11
Safety Precautions.. 22
Measuring Voltage..33
Measuring Current..46
Measuring Resistance...58
Advanced Functions... 71
Troubleshooting Techniques....................................85
Practical Applications.. 99
Maintenance and Calibration.................................112

Introduction to Multimeters

History and Evolution

The multimeter, a quintessential tool in the field of electronics and electrical engineering, has a rich history that traces back to the early 20th century. The invention of the first multimeter is credited to British Post Office engineer Donald Macadie in 1923. Frustrated with the need to carry multiple instruments for electrical measurements, Macadie invented a device that could measure voltage, current, and resistance, naming it the "Avometer." This analog multimeter utilised moving coil metres to provide readings.

Over the decades, the multimeter underwent significant transformations. The 1960s and 1970s saw the advent of solid-state electronics, which led to the development of digital multimeters (DMMs). These devices replaced analog dials with digital displays, enhancing precision and ease of use. Today, multimeters are indispensable in various fields, from simple home electrical repairs to complex industrial applications, owing to their versatility and accuracy.

Types of Multimeters
There are two primary types of multimeters: analog and digital.

Analog Multimeters:
Analog multimeters use a microammeter with a moving pointer to display readings. They are favoured for their ability to show trends and variations in the measurements. While they might lack the precision of their digital counterparts, analog multimeters are still in use due to their robustness and the intuitive nature of their readouts.

Digital Multimeters (DMMs):
Digital multimeters have largely replaced analog multimeters in modern applications. They use an analog-to-digital converter to display

measurements on an LCD or LED screen. DMMs offer higher accuracy, greater ease of use, and additional features such as auto-ranging, data logging, and connectivity to computers for advanced analysis.

Components of a Multimeter

Understanding the various parts of a multimeter is crucial for its effective use. Here are the key components:

Display: The display is where the measurement readings are shown. Digital multimeters have an LCD or LED screen, while analog multimeters have a dial with a moving needle.

Selector Dial: Also known as the rotary switch, this dial allows the user to select the type of measurement (voltage, current, resistance, etc.) and the appropriate range.

Test Leads: These are insulated wires with metal probes at the ends, used to connect the multimeter to the circuit or component being tested. Typically, the red lead is positive, and the black lead is negative.

Input Terminals: These are the ports where the test leads are inserted. Common terminals include COM (common ground), VΩ (for voltage and resistance), and mA/10A (for measuring current).

Symbols and Units
Multimeters use various symbols and units to indicate different measurements. Understanding these symbols is essential for correct usage:

V (Voltage): Measured in volts (V). Voltage can be AC (alternating current) or DC (direct current), usually denoted by a tilde (~) for AC and a straight line with a dashed line underneath (— —) for DC.

A (Current): Measured in amperes (A). Similar to voltage, current can be AC or DC.

Ω (Resistance): Measured in ohms (Ω), representing the opposition to the flow of current.

m (Milli): Denotes one-thousandth (0.001). For example, 1 mV (millivolt) is 0.001 volts.

μ (Micro): Denotes one-millionth (0.000001). For example, 1 μA (microampere) is 0.000001 amperes.

Basic Principles of Operation

A multimeter operates on the principle of measuring the voltage drop across a known resistance and using Ohm's Law to calculate the desired parameter. Ohm's Law states that $V = IR$, where V is voltage, I is current, and R is resistance. By rearranging this equation, a multimeter can determine one of the variables if the other two are known.

Applications of Multimeters

Multimeters have a wide range of applications, making them invaluable tools in various settings:

Home Use: For checking batteries, troubleshooting electrical outlets, and verifying the continuity of wiring.

Automotive: Used for diagnosing electrical problems in vehicles, such as checking the battery voltage, testing fuses, and verifying the functionality of sensors.

Industrial: Essential for maintenance and troubleshooting of industrial equipment, including motors, control systems, and power distribution networks.

Electronics: Crucial for designing, testing, and repairing electronic circuits and components, such as resistors, capacitors, and transistors.

Selecting the Right Multimeter
Choosing the right multimeter depends on the intended use and the specific requirements of the user:

Accuracy: Higher accuracy is crucial for precise measurements, particularly in professional and industrial applications.

Range: Ensure the multimeter can measure the expected ranges of voltage, current, and resistance for your applications.

Features: Consider additional features such as auto-ranging, data logging, connectivity options, and safety ratings.

Build Quality: A robust and durable multimeter is essential for long-term use, especially in demanding environments.

Understanding the Basics

Parts of a Multimeter
A multimeter, also known as a volt-ohm metre (VOM), is an essential tool for anyone working with electrical circuits. It combines several measurement functions into one unit, including voltage, current, and resistance. Understanding the basic parts of a multimeter is crucial for effective and safe use.

Display:
The display is where the measured values are shown. Digital multimeters (DMMs) have an LCD or LED screen that provides a numerical readout, which is easy to read and typically more accurate than an analog display. Some advanced DMMs feature a backlit display for use in low-light conditions.

Selector Dial:

The selector dial, also known as the rotary switch, allows the user to choose the type of measurement and the range. This dial can be turned to various positions to select functions such as AC voltage (V~), DC voltage (V—), AC current (A~), DC current (A—), resistance (Ω), and other functions like capacitance, frequency, and continuity testing.

Test Leads:

Test leads are insulated wires that connect the multimeter to the circuit or component being tested. They usually come in a pair: red for the positive (live) connection and black for the negative (ground) connection. The leads are equipped with metal probes, alligator clips, or specialised adapters to facilitate connections.

Input Terminals:

Input terminals are the ports where the test leads are plugged into the multimeter. Common terminals include:

- COM (Common): The black lead is typically inserted here. It acts as the ground or reference point.
- VΩ (Voltage/Resistance): The red lead is inserted here for measuring voltage and resistance.
- mA (Milliampere): Used for measuring small currents.
- 10A: Used for measuring larger currents. This terminal is usually fused to protect the multimeter from high currents.

Function Buttons:
Advanced multimeters come with additional function buttons that enhance their usability. These buttons may include:

- Hold: Freezes the current reading on the display, allowing for easier recording.
- Min/Max: Captures and displays the minimum and maximum values measured over a period.

- Range: Allows manual selection of the measurement range, offering more control over the measurement process.
- Mode: Switches between different modes within a measurement type (e.g., AC/DC for voltage or current).

Symbols and Units

Multimeters use various symbols and units to indicate different measurements. Recognizing and understanding these symbols is vital for correct usage:

Voltage (V):
Voltage is measured in volts (V). Multimeters can measure both alternating current (AC) and direct current (DC) voltages. AC voltage is denoted by a tilde (~), while DC voltage is represented by a straight line with a dashed line underneath (— —).

Current (A):
Current is measured in amperes (A). Similar to voltage, current can be AC or DC. AC current is

indicated by a tilde (~), and DC current is shown by a straight line with a dashed line underneath (– –).

Resistance (Ω):
Resistance is measured in ohms (Ω), representing the opposition to the flow of current. The symbol for resistance is the Greek letter omega (Ω).

Capacitance (F):
Capacitance is measured in farads (F), indicating the ability of a component to store an electrical charge. Capacitance is often measured in smaller units such as microfarads (μF) or nanofarads (nF).

Frequency (Hz):
Frequency is measured in hertz (Hz) and indicates the number of cycles per second in an AC signal. This measurement is particularly important in applications involving signal processing and communications.

Continuity:

Continuity is checked to determine if there is a complete path for current flow in a circuit. It is often indicated by a diode symbol or a sound wave symbol. When continuity is detected, the multimeter usually emits a beep, signifying a closed circuit.

Diode Test:

This function tests the voltage drop across a diode. A healthy diode will show a voltage drop when forward-biassed and an open circuit when reverse-biassed. The diode test mode is usually indicated by a diode symbol.

Measuring Techniques

To effectively use a multimeter, it's essential to know how to measure different parameters:

Measuring Voltage:

- Set the Selector Dial: Turn the dial to the appropriate voltage type (AC or DC) and range.

- Connect the Test Leads: Insert the black lead into the COM terminal and the red lead into the VΩ terminal.
- Probe the Circuit: Place the test leads across the component or circuit where you need to measure voltage. Ensure proper polarity for DC measurements.
- Read the Display: The voltage reading will appear on the display. For AC measurements, the display will show the RMS (root mean square) value of the voltage.

Measuring Current:

- Set the Selector Dial: Turn the dial to the appropriate current type (AC or DC) and range.
- Connect the Test Leads: Insert the black lead into the COM terminal and the red lead into the appropriate current terminal (mA or 10A) depending on the expected current.

- Break the Circuit: Unlike voltage measurement, current measurement requires placing the multimeter in series with the circuit. Disconnect the circuit at the point where you want to measure current and connect the test leads in series.
- Read the Display: The current reading will appear on the display.

Measuring Resistance:

- Set the Selector Dial: Turn the dial to the resistance (Ω) setting.
- Connect the Test Leads: Insert the black lead into the COM terminal and the red lead into the VΩ terminal.
- Probe the Component: Place the test leads across the component whose resistance you want to measure. Ensure the component is not connected to any power source.
- Read the Display: The resistance value will appear on the display. For accurate

measurements, ensure the component is isolated from the circuit.

Practical Tips for Accurate Measurements
Auto-Ranging vs. Manual Ranging:

- Auto-Ranging: Many modern multimeters have an auto-ranging feature that automatically selects the appropriate range for the measurement. This simplifies the process and reduces the chance of errors.
- Manual Ranging: If your multimeter requires manual ranging, start with the highest range and work your way down to prevent overloading the metre.

Zeroing the Multimeter:
Before measuring resistance, touch the test leads together and ensure the multimeter reads zero or near zero. This step compensates for the resistance of the leads themselves.

Battery Check:

Ensure the multimeter's battery is in good condition. A low battery can cause inaccurate readings.

Temperature Considerations:
Be aware that extreme temperatures can affect the accuracy of measurements. Try to perform measurements at room temperature whenever possible.

Reading Stability:
Allow a few seconds for the reading to stabilise before recording it, especially for measurements involving high resistance or low current.

Safety Precautions

When using a multimeter, it's crucial to follow safety precautions to protect yourself, your equipment, and the electrical systems you are working on.

Personal Safety Tips

Understand the Tool:
Before using a multimeter, thoroughly read the user manual to understand its functions, limitations, and safety warnings. Different models may have specific instructions and safety features.

Proper Attire:
Wear appropriate clothing and protective gear. Avoid loose clothing or jewellery that could

accidentally come into contact with electrical components. Insulated gloves and safety glasses are recommended when working with high-voltage or high-current circuits.

Dry Hands:
Ensure your hands are dry before handling the multimeter or any electrical components. Moisture can increase the risk of electrical shock.

One-Hand Rule:
When working with high voltage, try to use only one hand. Keep your other hand away from all conductive materials. This reduces the chance of an electrical current passing through your heart in case of accidental contact.

Equipment Safety Tips

Inspect the Multimeter:
Before each use, inspect the multimeter for any physical damage. Check the casing, display, selector dial, and input terminals for cracks or

signs of wear. Ensure the test leads are in good condition without any exposed wires or damaged insulation.

Use Proper Test Leads:
Always use the test leads provided with the multimeter or certified replacements. Ensure they are rated for the voltages and currents you will be measuring. Using inferior or damaged leads can result in inaccurate measurements or electrical hazards.

Correct Input Terminals:
Always plug the test leads into the correct input terminals. For example, use the COM terminal for the common (black) lead and the VΩ terminal for voltage and resistance measurements. Using the wrong terminals can damage the multimeter and pose a safety risk.

Check the Multimeter's Rating:
Ensure your multimeter is rated for the electrical environment you are working in. Look for the CAT (Category) rating:

- CAT I: For protected electronic equipment.
- CAT II: For household appliances and portable tools.
- CAT III: For distribution boards, circuit breakers, and wiring.
- CAT IV: For utility connections and outdoor conductors.

Fused Inputs:
Use multimeters with fused input terminals for current measurements. The fuse will protect the multimeter and the user in case of an accidental overload.

Avoid Overloading:
Do not attempt to measure a parameter that exceeds the multimeter's specified limits. Overloading can cause damage to the multimeter and pose serious safety risks.

Disconnect Power:

Whenever possible, disconnect the power source before connecting or disconnecting the multimeter. This minimises the risk of short circuits and accidental contact with live parts.

Measuring Voltage
Verify Settings:
Always double-check that the multimeter is set to the correct mode and range for the measurement you intend to take. Incorrect settings can lead to inaccurate readings or damage to the multimeter.

Secure Connections:
Ensure that the test leads are securely connected to the multimeter and the circuit. Loose connections can lead to arcing, which poses a fire hazard.

Probing:
Be careful when probing circuits, especially in crowded or confined spaces. Avoid touching the metal part of the test leads to prevent accidental

shocks. Use one probe at a time and avoid bridging across components.

Isolation:
Isolate the circuit under test as much as possible. This reduces the risk of accidental contact with other live parts of the circuit.

Measuring Current

Series Connection:
When measuring current, always connect the multimeter in series with the circuit. Do not connect it across a voltage source as this can cause a short circuit and damage the multimeter.

Expected Current:
Estimate the expected current before measuring. Start with the highest current range on the multimeter and work your way down to prevent overloading the metre.

Replace Fuses:

If the multimeter's fuse blows, replace it with one of the same rating. Do not use a fuse with a higher rating or bypass the fuse as this negates the protective function and can lead to damage or injury.

Measuring Resistance

Power Off:
Ensure the circuit is powered off and fully discharged before measuring resistance. Measuring resistance in a live circuit can damage the multimeter and result in incorrect readings.

Component Isolation:
Isolate the component from the circuit before measuring its resistance. This prevents other parallel paths from affecting the measurement and provides an accurate reading.

Capacitor Discharge:
If you are measuring resistance in a circuit that contains capacitors, ensure they are fully

discharged to avoid damaging the multimeter or getting an inaccurate reading.

Advanced Functions

Capacitance and Frequency:
When measuring capacitance or frequency, ensure you are using a multimeter designed for those measurements. Incorrect usage can lead to inaccurate readings or damage to the multimeter.

Diode Testing:
Use the diode test function only on unpowered circuits. Applying voltage to a circuit in diode test mode can damage the multimeter and the diode being tested.

Continuity Testing:
When using the continuity function, ensure the circuit is unpowered. Continuity testing in a live circuit can damage the multimeter and provide inaccurate results.

General Safety Tips

Work Area:
Keep your work area clean and organised. A cluttered workspace can lead to accidents and make it harder to follow safety procedures.

Emergency Procedures:
Be familiar with emergency procedures, such as shutting off power and using a fire extinguisher. Know the location of emergency shut-off switches and fire extinguishers in your work area.

Training:
Ensure you are properly trained in using the multimeter and understand the specific risks associated with the electrical systems you are working on. Regularly update your knowledge and skills to stay safe.

Follow Local Codes:
Adhere to local electrical codes and standards. These regulations are designed to ensure safety and should be followed rigorously.

Regular Maintenance:
Perform regular maintenance checks on your multimeter, including battery replacement and calibration. A well-maintained multimeter provides more accurate readings and reduces the risk of malfunction.

Avoid Static Discharge:
Be aware of static discharge, especially when working with sensitive electronic components. Use anti-static mats and wrist straps to protect both the multimeter and the components.

Measuring Voltage

Measuring voltage is one of the primary functions of a multimeter. Whether you're troubleshooting a circuit, verifying power supply levels, or testing batteries, accurate voltage measurements are essential. This section will cover the types of voltage measurements, the steps involved in measuring voltage with a multimeter, and best practices to ensure accurate and safe measurements.

Types of Voltage
Direct Current (DC) Voltage:
DC voltage is a steady, constant voltage typically supplied by batteries and DC power supplies. It flows in one direction and is used in most electronic devices and automotive applications.

Alternating Current (AC) Voltage:
AC voltage periodically reverses direction and varies in amplitude over time. It is commonly found in household and industrial power supplies, with the standard frequency being 50 or 60 Hz depending on the region.

Steps for Measuring Voltage

Select the Appropriate Voltage Type and Range:

- Turn the selector dial on your multimeter to the DC voltage (V—) or AC voltage (V~) setting, depending on the type of voltage you are measuring.
- If your multimeter has manual ranging, select a voltage range that is higher than the expected voltage. If unsure, start with the highest range and work your way down.

Connect the Test Leads:

- Insert the black test lead into the COM (common) terminal.
- Insert the red test lead into the VΩ terminal. Ensure the leads are fully inserted and secure.

Check the Multimeter:

- Verify that the multimeter is functioning correctly and that the battery is not low. A low battery can lead to inaccurate readings.

Prepare the Circuit:

- Ensure that the circuit or component to be tested is accessible and safe to probe. If you are measuring the voltage of a power supply or battery, make sure you can safely access the terminals.

Place the Probes:

- Touch the black probe to the negative or ground point of the circuit.
- Touch the red probe to the positive point of the circuit.
- For DC measurements, ensure the correct polarity. For AC measurements, polarity does not matter as the voltage alternates.

Read the Display:

- Observe the reading on the multimeter's display. If you are using a digital multimeter, the voltage will be shown as a numerical value.
- If the reading is negative in a DC measurement, it indicates that the probes are reversed.

Interpret the Results:

- Compare the measured voltage with the expected value to determine if the circuit or component is functioning correctly. Significant deviations might indicate issues such as faulty components, bad connections, or power supply problems.

Turn Off the Multimeter:

- After taking the measurement, turn off the multimeter to conserve battery life and disconnect the test leads.

Best Practices for Accurate Voltage Measurements

Stability:
Ensure that the circuit is stable and not fluctuating during the measurement. This is especially important for AC voltage, where fluctuations can affect the accuracy of the reading.

Proper Connections:
Make sure the test leads are securely connected to both the multimeter and the circuit under test. Loose connections can lead to inaccurate readings or intermittent results.

Auto-Ranging:
If your multimeter has an auto-ranging feature, use it to automatically select the appropriate range for the measurement. This simplifies the process and reduces the chance of selecting an incorrect range.

Calibration:
Regularly calibrate your multimeter to ensure it provides accurate readings. Calibration should be performed according to the manufacturer's recommendations.

Environmental Factors:
Be aware of environmental factors such as temperature, humidity, and electromagnetic interference, which can affect the accuracy of voltage measurements. Whenever possible, take measurements in a controlled environment.

Zero Adjustment:
For analog multimeters, perform a zero adjustment before taking measurements to

ensure accuracy. This involves setting the needle to zero when the probes are shorted together.

Measuring Specific Types of Voltage
Measuring Battery Voltage:

- Set the multimeter to DC voltage.
- Place the black probe on the negative terminal and the red probe on the positive terminal of the battery.
- Read the voltage. A fully charged standard AA battery, for example, should read around 1.5V.

Measuring Power Supply Voltage:

- Set the multimeter to the appropriate AC or DC voltage setting.
- Connect the black probe to the ground or negative terminal.
- Connect the red probe to the positive terminal or output terminal of the power supply.

- Read the voltage to ensure it matches the expected output of the power supply.

Measuring Mains Voltage:

- Set the multimeter to AC voltage.
- Connect the black probe to the ground or neutral point.
- Connect the red probe to the live terminal.
- Read the voltage. Standard household voltage should be around 120V or 230V, depending on the country.
- Use extreme caution when measuring mains voltage due to the high risk of electric shock.

Measuring Voltage Drops:

- Set the multimeter to the appropriate voltage setting (AC or DC).
- Measure the voltage across the component or section of the circuit where the drop is expected.

- Compare the voltage drop to the expected value to identify potential issues like poor connections or faulty components.

Troubleshooting Voltage Measurements

Inconsistent Readings:

- Ensure stable connections and probe placement.
- Check the condition of the test leads and replace them if damaged.
- Verify that the circuit is stable and not fluctuating excessively.

No Reading:

- Confirm that the multimeter is set to the correct mode and range.
- Ensure that the circuit is powered and that the probes are making proper contact.
- Check the multimeter's battery and replace it if necessary.

Overload Indication:

- Switch to a higher voltage range if your multimeter displays an overload or out-of-range indication.
- Verify the expected voltage before measuring to avoid overloading the multimeter.

Negative Reading:

- In DC voltage measurements, a negative reading indicates reversed polarity. Swap the probe connections to correct this.

Fluctuating Readings:

- Check for loose or intermittent connections.
- Ensure that the test leads are not moving during the measurement.
- Consider environmental factors such as electromagnetic interference.

Safety Precautions

High Voltage:

- When measuring high voltage, use insulated gloves and stand on an insulated mat to prevent electric shock.
- Keep your fingers behind the finger guards on the test probes.

Live Circuits:

- Exercise caution when probing live circuits to avoid accidental contact with live parts.
- Use one hand to probe and keep the other hand away from the circuit to reduce the risk of current passing through your body.

Capacitors:

- Discharge capacitors before measuring voltage to prevent electric shock and damage to the multimeter.

Proper Storage:

- Store your multimeter and test leads in a dry, dust-free environment to maintain their condition and accuracy.

Regular Inspection:

Periodically inspect the multimeter and test leads for any signs of wear or damage. Replace any faulty components immediately.

Measuring Current

Measuring current is a critical function of a multimeter, allowing you to diagnose issues, verify operation, and ensure the safety and efficiency of electrical circuits. This section covers the types of current, steps for measuring current using a multimeter, and best practices for accurate and safe measurements.

Types of Current

Direct Current (DC):
DC current flows in one direction and is typically found in battery-powered devices and electronic circuits. It is steady and constant over time.

Alternating Current (AC):
AC current periodically reverses direction and is commonly used in household and industrial

power supplies. The frequency of AC current is usually 50 or 60 Hz, depending on the region.

Steps for Measuring Current

Set Up the Multimeter:

- Turn the selector dial to the appropriate current measurement setting (AC or DC) and range. If your multimeter has an auto-ranging feature, set it to the current mode, and it will automatically select the appropriate range.
- Insert the black test lead into the COM (common) terminal.
- Insert the red test lead into the appropriate current terminal (mA or 10A) depending on the expected current. Use the 10A terminal for higher currents to prevent blowing the fuse in the mA terminal.

Break the Circuit:

Unlike voltage measurement, measuring current requires the multimeter to be placed in series with the circuit. This means you need to open the circuit at the point where you want to measure the current.

Connect the Multimeter in Series:

- Connect the multimeter probes to the two ends of the break in the circuit. The current will flow through the multimeter, allowing it to measure the flow accurately.
- Ensure that the connections are secure and that the probes are firmly attached to the circuit.

Power On the Circuit:

- Once the multimeter is connected in series, power on the circuit. The current will flow through the multimeter, and the reading will appear on the display.

Read the Display:

- Observe the reading on the multimeter display. If using a digital multimeter, the current will be shown as a numerical value in amperes (A), milliamperes (mA), or microamperes (μA).

Turn Off the Circuit:

- After taking the measurement, turn off the circuit and remove the multimeter probes. Reconnect the circuit to restore normal operation.
- Best Practices for Accurate Current Measurements

Estimate the Current:
Before measuring, estimate the expected current to select the appropriate range. Starting with the highest range available on the multimeter helps prevent overloading the metre.

Use Proper Terminals:

- Always use the correct input terminal for the current range. For high currents, use the 10A terminal. For lower currents, use the mA or μA terminal. Using the wrong terminal can damage the multimeter and the circuit.

Check Fuse Ratings:
Ensure that the fuse in the multimeter's current measurement terminal is of the correct rating. If the fuse blows, replace it with one of the same rating to maintain safety and functionality.

Avoid Overloading:
Never attempt to measure current that exceeds the multimeter's specified limits. Overloading can cause internal damage and pose a safety risk.

Secure Connections:
Ensure that all connections are secure and stable. Loose connections can lead to inaccurate readings or damage the multimeter.

Use High-Quality Test Leads:

High-quality test leads with proper insulation and connectors ensure accurate measurements and reduce the risk of accidental shorts or shocks.

Minimise Interference:
Keep the test leads and multimeter away from sources of electromagnetic interference, which can affect the accuracy of current measurements.

Measuring Specific Types of Current

Measuring DC Current:

- Set the multimeter to the DC current (A—) setting.
- Break the circuit at the point where the current is to be measured.
- Connect the multimeter in series with the circuit.
- Power on the circuit and read the current value on the display.

Measuring AC Current:

- Set the multimeter to the AC current (A~) setting.
- Break the circuit at the point where the current is to be measured.
- Connect the multimeter in series with the circuit.
- Power on the circuit and read the current value on the display.

Measuring High Current:

- Set the multimeter to the appropriate high current setting (usually 10A).
- Break the circuit and connect the multimeter in series.
- Ensure the connections are secure to handle the high current.
- Power on the circuit and read the current value on the display.
- For very high currents, consider using a clamp metre, which can measure current without breaking the circuit.

Troubleshooting Current Measurements

No Reading:

- Ensure the multimeter is set to the correct current mode and range.
- Verify that the circuit is powered and that the probes are making proper contact.
- Check for a blown fuse in the multimeter and replace it if necessary.

Inconsistent Readings:

- Ensure stable connections and secure probe placement.
- Check the condition of the test leads and replace them if damaged.
- Verify that the circuit is stable and not fluctuating excessively.

Overload Indication:

- Switch to a higher current range if your multimeter displays an overload or out-of-range indication.
- Verify the expected current before measuring to avoid overloading the multimeter.

Fluctuating Readings:

- Check for loose or intermittent connections.
- Ensure that the test leads are not moving during the measurement.
- Consider environmental factors such as electromagnetic interference.

Safety Precautions

High Current:

- When measuring high current, use insulated gloves and stand on an insulated mat to prevent electric shock.

- Keep your fingers behind the finger guards on the test probes.

Live Circuits:

- Exercise caution when probing live circuits to avoid accidental contact with live parts.
- Use one hand to probe and keep the other hand away from the circuit to reduce the risk of current passing through your body.

Fuse Protection:

- Ensure the multimeter's current measurement terminals are fused to protect against accidental overloading.
- Replace blown fuses with ones of the same rating to maintain safety.

Proper Storage:

- Store your multimeter and test leads in a dry, dust-free environment to maintain their condition and accuracy.

Regular Inspection:

- Periodically inspect the multimeter and test leads for any signs of wear or damage. Replace any faulty components immediately.

Avoid Static Discharge:

Be aware of static discharge, especially when working with sensitive electronic components. Use anti-static mats and wrist straps to protect both the multimeter and the components.

Measuring Resistance

Measuring resistance is a fundamental task for diagnosing and troubleshooting electrical and electronic circuits. Resistance measurement allows you to check the integrity of components, connections, and circuits. This section covers the principles of resistance, steps for measuring resistance using a multimeter, and best practices for accurate and safe measurements.

Principles of Resistance

Definition of Resistance:
Resistance is the opposition to the flow of electric current in a circuit. It is measured in ohms (Ω). Higher resistance means less current flow, while lower resistance means more current flow.

Ohm's Law:
Ohm's Law describes the relationship between voltage (V), current (I), and resistance (R) in a circuit:

$$V = I \times R$$

V=I×R. This fundamental principle helps in understanding how resistance affects current and voltage.

Resistors:
Resistors are components specifically designed to provide a precise amount of resistance in a circuit. They are used to control current flow, divide voltages, and protect components.

Factors Affecting Resistance:

- Material: Different materials have different resistivities. Conductors have

low resistance, while insulators have high resistance.
- Temperature: Resistance can change with temperature. For most conductors, resistance increases with temperature.
- Length and Cross-Sectional Area: Longer conductors have higher resistance, while conductors with larger cross-sectional areas have lower resistance.

Steps for Measuring Resistance

Prepare the Multimeter:

- Turn the selector dial to the resistance (Ω) setting. If your multimeter has manual ranging, select the appropriate range for the expected resistance.

Check the Multimeter:

- Ensure the multimeter is functioning correctly and that the battery is not low. A low battery can lead to inaccurate readings.

Disconnect Power:

- Ensure the circuit or component to be tested is powered off and fully discharged. Measuring resistance in a live circuit can damage the multimeter and result in incorrect readings.

Isolate the Component:

- For accurate measurements, isolate the component from the circuit. This prevents parallel paths from affecting the measurement.

Connect the Test Leads:

- Insert the black test lead into the COM (common) terminal.

- Insert the red test lead into the VΩ terminal. Ensure the leads are fully inserted and secure.

Place the Probes:

- Touch the black probe to one end of the component or circuit.
- Touch the red probe to the other end.

Read the Display:

- Observe the reading on the multimeter's display. If using a digital multimeter, the resistance will be shown as a numerical value in ohms (Ω), kilohms (kΩ), or megohms (MΩ).

Interpret the Results:

- Compare the measured resistance with the expected value to determine if the component or circuit is functioning correctly. Significant deviations might indicate issues such as damaged components or poor connections.

Turn Off the Multimeter:

- After taking the measurement, turn off the multimeter to conserve battery life and disconnect the test leads.

Best Practices for Accurate Resistance Measurements

Stable Connections:
Ensure that the test leads are securely connected to both the multimeter and the component under test. Loose connections can lead to inaccurate readings or intermittent results.

Zero Adjustment:
For analog multimeters, perform a zero adjustment before taking measurements to ensure accuracy. This involves setting the needle to zero when the probes are shorted together.

Temperature Considerations:
Be aware of temperature effects on resistance. Measure in a stable temperature environment, and consider the temperature coefficient of the component being tested.

Clean Contacts:
Ensure that the contact points are clean and free from oxidation or contaminants. Dirty contacts can introduce additional resistance and affect the accuracy of the measurement.

Component Isolation:
Isolate the component from the circuit to avoid parallel paths that can affect the measurement. This is particularly important for low-resistance measurements.

Measuring Specific Types of Resistance

Measuring Resistors:

- Set the multimeter to the appropriate resistance range.

- Connect the probes to the resistor leads.
- Read the resistance value on the display and compare it with the resistor's marked value or tolerance.

Measuring Continuity:

- Set the multimeter to the continuity test mode (often indicated by a sound wave symbol).
- Connect the probes to the two points in the circuit.
- A continuous circuit will cause the multimeter to beep, indicating low resistance and good continuity.

Measuring Potentiometers:

Set the multimeter to the appropriate resistance range.
Connect the probes to the wiper (middle terminal) and one of the outer terminals.
Rotate the potentiometer knob and observe the change in resistance on the display.

Measuring Wire Resistance:

Set the multimeter to a low resistance range. Connect the probes to both ends of the wire. Read the resistance value. For long wires, expect a small resistance value proportional to the length and gauge of the wire.

Measuring Insulation Resistance:

- Set the multimeter to the highest resistance range.
- Connect the probes to the insulation material.
- Read the resistance value. Insulation should show very high resistance, typically in the megohms (MΩ) range.

Troubleshooting Resistance Measurements

No Reading or Overload:

- Ensure the multimeter is set to the correct resistance mode and range.

- Verify that the component or circuit is properly isolated and not powered.
- Check for a damaged component or broken circuit that might result in infinite resistance.

Inconsistent Readings:

- Ensure stable connections and secure probe placement.
- Check the condition of the test leads and replace them if damaged.
- Verify that the contact points are clean and free from oxidation.

Negative or Erratic Readings:

- Ensure that the multimeter is functioning correctly and that the battery is not low.
- Check for electromagnetic interference that might affect the readings.
- Inspect the component for any signs of damage or malfunction.

Low Resistance Measurements:

- Use the lowest resistance range on the multimeter for accurate low-resistance measurements.
- Ensure that the contact points are clean and that the test leads are in good condition.
- Short the probes together before measurement to check for and subtract any residual resistance of the leads.

Safety Precautions

Power Off:
Always ensure the circuit is powered off and fully discharged before measuring resistance. Measuring in a live circuit can damage the multimeter and result in inaccurate readings.

Component Handling:
Handle components carefully to avoid damage. Some components, like semiconductors, can be sensitive to static discharge.

Proper Storage:
Store your multimeter and test leads in a dry, dust-free environment to maintain their condition and accuracy.

Regular Inspection:
Periodically inspect the multimeter and test leads for any signs of wear or damage. Replace any faulty components immediately.

Avoid Static Discharge:
Be aware of static discharge, especially when working with sensitive electronic components. Use anti-static mats and wrist straps to protect both the multimeter and the components.

Advanced Functions

Modern multimeters are equipped with a variety of advanced functions that go beyond basic voltage, current, and resistance measurements. These advanced features provide enhanced diagnostic capabilities, making multimeters versatile tools for a wide range of applications.

Capacitance Measurement

Understanding Capacitance:
Capacitance is the ability of a component to store an electrical charge. It is measured in farads (F), with common units being microfarads (µF) and picofarads (pF).

Steps to Measure Capacitance:

- Set the multimeter to the capacitance (F) mode.

- Ensure the capacitor is fully discharged before measuring.
- Connect the multimeter probes to the capacitor terminals (polarity usually does not matter for non-polarized capacitors).
- Read the capacitance value on the display.

Applications:

- Checking capacitors in power supplies and audio equipment.
- Verifying the value of capacitors in electronic circuits.
- Diagnosing capacitor failures, which can cause issues like power supply instability or audio distortion.

Frequency and Duty Cycle Measurement

Understanding Frequency and Duty Cycle:

- Frequency: The number of cycles per second of an alternating current (AC) signal, measured in hertz (Hz).

- Duty Cycle: The percentage of one period in which a signal is active. It is a crucial parameter for pulse-width modulation (PWM) signals.

Steps to Measure Frequency:

- Set the multimeter to the frequency (Hz) mode.
- Connect the multimeter probes across the signal source.
- Read the frequency value on the display.

Steps to Measure Duty Cycle:

- Set the multimeter to the duty cycle (%) mode.
- Connect the multimeter probes across the signal source.
- Read the duty cycle percentage on the display.

Applications:

- Verifying the frequency of oscillators and clock signals in digital circuits.
- Checking the duty cycle of PWM signals used in motor control and dimming applications.
- Diagnosing timing issues in electronic circuits.

Temperature Measurement

Understanding Temperature Measurement: Some multimeters come with a thermocouple probe to measure temperature. The probe converts temperature into an electrical signal, which the multimeter reads and displays.

Steps to Measure Temperature:

- Set the multimeter to the temperature (°C or °F) mode.
- Connect the thermocouple to the multimeter.

- Place the thermocouple probe in contact with the object or environment being measured.
- Read the temperature value on the display.

Applications:

- Monitoring the temperature of components and circuits to prevent overheating.
- Measuring ambient temperature in HVAC systems.
- Checking the temperature of soldering irons and other tools.

Continuity Testing

Understanding Continuity Testing:
Continuity testing checks if there is a complete path for current to flow between two points in a circuit. It is indicated by a beeping sound or a visual indicator on the multimeter.

Steps to Perform Continuity Testing:

- Set the multimeter to the continuity test mode (usually indicated by a sound wave symbol).
- Connect the multimeter probes to the two points being tested.
- Listen for the beep or look for the visual indicator, which signifies continuity (low resistance path).

Applications:

- Checking connections and traces on PCBs.
- Verifying the integrity of wires and cables.
- Testing switches and relays for proper operation.

Diode Testing

Understanding Diode Testing:
The diode test function checks the forward voltage drop of a diode, indicating if the diode is functioning correctly. Diodes typically have a

forward voltage drop of 0.6 to 0.7 volts for silicon diodes and 0.2 to 0.3 volts for germanium diodes.

Steps to Perform Diode Testing:

- Set the multimeter to the diode test mode (usually indicated by a diode symbol).
- Connect the red probe to the anode and the black probe to the cathode of the diode.
- Read the forward voltage drop on the display. A typical reading should be within the expected range for the diode type.
- Reverse the probes to check for reverse bias, which should show no conduction (infinite resistance).

Applications:

- Checking the health of diodes in power supplies and rectifiers.

- Testing LEDs and other semiconductor devices.
- Diagnosing issues in circuits that use diodes for protection or signal processing.

True RMS Measurement

Understanding True RMS:
True RMS (Root Mean Square) measurement is essential for accurately measuring AC signals that are not pure sine waves, such as those found in variable frequency drives (VFDs) and other complex waveforms.

Steps to Use True RMS:

- Set the multimeter to the AC voltage or current mode with True RMS capability.
- Connect the multimeter probes to the circuit as needed.
- Read the True RMS value on the display.

Applications:

- Measuring AC signals with complex waveforms.
- Verifying the output of inverters and VFDs.
- Ensuring accurate readings in circuits with non-linear loads.

Data Logging

Understanding Data Logging:
Data logging allows a multimeter to record measurements over time, which can be useful for monitoring changes in electrical parameters or diagnosing intermittent issues.

Steps to Use Data Logging:

- Set up the multimeter with the data logging function enabled.
- Configure the logging parameters, such as interval and duration.
- Start the logging process and let the multimeter record the data.

- Retrieve and analyse the recorded data from the multimeter's memory or connect it to a computer for further analysis.

Applications:

Monitoring power supply stability over time.
- Recording temperature variations in environmental testing.
- Diagnosing intermittent faults in electrical systems.

Auto-Hold and Min/Max Functions

Understanding Auto-Hold and Min/Max:

- Auto-Hold: This function freezes the displayed value automatically when the measurement stabilises, making it easier to capture and read measurements without looking at the display constantly.
- Min/Max: This function records the minimum and maximum values of a

parameter over time, helping to identify fluctuations.

Steps to Use Auto-Hold:

- Enable the auto-hold function on the multimeter.
- Perform the measurement as usual.
- The multimeter will automatically freeze the stable value on the display.

Steps to Use Min/Max:

- Enable the Min/Max function on the multimeter.
- Perform the measurement and allow the multimeter to record values over time.
- Review the recorded minimum and maximum values to understand the range of fluctuations.

Applications:

- Capturing transient events or spikes in electrical signals.
- Monitoring variations in environmental conditions.
- Simplifying the measurement process in dynamic systems.

Relative Mode

Understanding Relative Mode:
Relative mode allows you to zero out the multimeter to a reference value, making it easier to measure changes from that reference point.

Steps to Use Relative Mode:

- Perform an initial measurement to obtain the reference value.
- Enable the relative mode function on the multimeter.
- Subsequent measurements will display the difference from the reference value.

Applications:

- Measuring voltage drops across components in a circuit.
- Checking for deviations in resistance from a known value.
- Comparing current changes relative to a baseline measurement.

Troubleshooting Techniques

Effective troubleshooting is essential for diagnosing and resolving electrical problems. A multimeter is an invaluable tool in this process, allowing you to measure voltage, current, resistance, and other parameters to identify issues.

Identifying Common Electrical Problems
Understanding the common types of electrical problems is the first step in troubleshooting. Here are some frequent issues and their symptoms:

1. Open Circuits:

- Symptoms: No power, no signal, or a non-functional device.

- Causes: Broken wires, loose connections, or failed components.
- Detection: Use a continuity test to check for complete paths.

2. Short Circuits:

- Symptoms: Blown fuses, tripped breakers, or sparks.
- Causes: Insulation failures, crossed wires, or damaged components.
- Detection: Measure resistance between points to find unexpectedly low resistance.

3. Overloaded Circuits:

- Symptoms: Frequent breaker trips, dimming lights, or overheating.
- Causes: Too many devices on one circuit, faulty appliances.
- Detection: Measure current draw and compare with circuit capacity.

4. Voltage Drops:

- Symptoms: Dim lights, slow motors, or erratic device behaviour.
- Causes: High resistance connections, undersized wiring.
- Detection: Measure voltage at different points in the circuit.

5. Ground Faults:

- Symptoms: Tripped ground fault circuit interrupters (GFCIs), shocks.
- Causes: Faulty insulation, water ingress, damaged equipment.
- Detection: Use a GFCI tester or measure resistance to ground.

6. Intermittent Problems:

- Symptoms: Sporadic device failures, flickering lights.
- Causes: Loose connections, thermal expansion, faulty components.

- Detection: Observe and measure parameters over time, use data logging.

7. Noise and Interference:

- Symptoms: Static, hum, or erratic behaviour in electronic devices.
- Causes: Electromagnetic interference (EMI), poor shielding.
- Detection: Use a multimeter with frequency measurement, oscilloscopes for detailed analysis.

Using a Multimeter for Diagnostics

A multimeter is a versatile tool for diagnosing electrical problems. Here's how to use it effectively:

1. Visual Inspection:

- Step: Before using a multimeter, perform a thorough visual inspection of the circuit or device.

Multimeter Mastery Handbook

- Objective: Look for obvious issues like burnt components, loose connections, or broken wires.

2. Safety First:

- Step: Ensure the power is off before connecting the multimeter to any circuit to prevent shock or damage.
- Objective: Follow proper safety protocols to avoid accidents.

3. Measuring Voltage:

DC Voltage:
- Set the multimeter to the appropriate DC voltage range.
- Connect the probes across the component or section of the circuit.
- Example: Measure the voltage of a battery or power supply.

AC Voltage:

- Set the multimeter to the appropriate AC voltage range.
- Connect the probes across the outlet or component.
- Example: Measure the voltage of a wall outlet to check for proper supply.

4. Measuring Current:

DC Current:
- Set the multimeter to the appropriate DC current range.
- Break the circuit and connect the multimeter in series.
- Example: Measure the current draw of a motor or electronic device.

AC Current:
- Set the multimeter to the appropriate AC current range.
- Break the circuit and connect the multimeter in series.
- Example: Measure the current draw of an appliance.

5. Measuring Resistance:

- Step: Set the multimeter to the appropriate resistance range.
- Procedure: Connect the probes to both ends of the component or section of the circuit.
- Example: Measure the resistance of a resistor or the continuity of a wire.

6. Continuity Testing:

- Step: Set the multimeter to continuity mode.
- Procedure: Connect the probes across the component or section of the circuit.
- Objective: Listen for a beep indicating continuity.
- Example: Test a fuse or switch for proper operation.

7. Diode Testing:

- Step: Set the multimeter to diode test mode.
- Procedure: Connect the red probe to the anode and the black probe to the cathode.
- Objective: Read the forward voltage drop, and reverse the probes to check for reverse bias.
- Example: Test a diode or LED for functionality.

8. Capacitance Measurement:

- Step: Set the multimeter to capacitance mode.
- Procedure: Ensure the capacitor is discharged, then connect the probes to the capacitor terminals.
- Objective: Read the capacitance value.
- Example: Check capacitors in power supplies or audio equipment.

9. Frequency and Duty Cycle Measurement:

- Step: Set the multimeter to frequency or duty cycle mode.
- Procedure: Connect the probes across the signal source.
- Objective: Read the frequency or duty cycle percentage.
- Example: Verify oscillator frequencies or PWM signals.

10. Temperature Measurement:

- Step: Set the multimeter to temperature mode.
- Procedure: Connect a thermocouple and place the probe in contact with the measurement point.
- Objective: Read the temperature value.
- Example: Monitor component temperatures to prevent overheating.

Troubleshooting Techniques

1. Systematic Approach:

- Divide and Conquer: Break the circuit into smaller sections and test each section individually.
- Top-Down or Bottom-Up: Choose a logical starting point, either from the power source down or from the load up.

2. Verify Power Supply:

- Check Voltage: Measure the voltage at the power supply to ensure it's within specifications.
- Check Connections: Ensure all connections are secure and free from corrosion.

3. Check Continuity:

- Test Wires and Traces: Use the continuity test to check for breaks or shorts in wires and PCB traces.
- Test Switches and Relays: Verify that switches and relays are functioning correctly.

4. Measure Resistances:

- Component Values: Measure the resistance of resistors and compare with expected values.
- Connections: Check for high resistance in connections that should have low resistance.

5. Observe Waveforms:

- Frequency and Duty Cycle: Use the frequency and duty cycle functions to check the integrity of signal sources.
- Noise and Interference: Look for unexpected signals that might indicate interference or faulty components.

6. Test Components:

- Diodes and Transistors: Use the diode test function to check diodes and transistors.

- Capacitors: Measure capacitance to verify capacitor health and function.

7. Monitor Temperature:

- Component Overheating: Use the temperature function to check for overheating components.
- Environmental Conditions: Ensure that environmental conditions are within safe operating ranges.

8. Log Data:

- Record Measurements: Use data logging to monitor parameters over time.
- Analyse Trends: Look for trends or patterns that might indicate the source of the problem.

9. Cross-Reference Documentation:

- Schematics and Manuals: Use schematics and manuals to understand circuit operation and expected values.
- Component Datasheets: Refer to datasheets for component specifications and test procedures.

10. Isolate and Test:

- Component Isolation: Isolate components to test them individually without interference from the rest of the circuit.
- Substitute Known Good Components: Replace suspected faulty components with known good ones to verify issues.

Practical Applications

Multimeters are versatile tools that are essential for diagnosing issues and performing tasks in various fields. Understanding how to apply a multimeter effectively can streamline troubleshooting and repair processes.

Automotive Troubleshooting

1. Diagnosing Electrical Issues:

Battery and Charging System:

- Battery Voltage: Measure the battery voltage to check its charge level. A healthy car battery should read around 12.6 volts when fully charged. When the engine is running, the voltage should be

between 13.7 and 14.7 volts, indicating that the alternator is charging the battery.
- Charging System Test: Start the engine and measure the voltage at the battery terminals. A voltage significantly higher than the battery's resting voltage indicates that the alternator is functioning correctly. Conversely, a low voltage suggests a problem with the alternator or its connections.

Starter Motor:

- Voltage Drop Test: Measure the voltage drop across the starter motor while cranking the engine. A large voltage drop can indicate high resistance in the starter circuit, which could be caused by corroded connections or a faulty starter.

Fuses and Relays:

- Continuity Testing: Check for continuity across fuses and relays to ensure they are not blown or faulty. For relays, measure the voltage at the relay terminals when activated to verify proper operation.

Sensors and Switches:

- Testing Sensors: Use the multimeter to measure the output of various sensors, such as the oxygen sensor or coolant temperature sensor. Compare the readings with the expected values specified in the vehicle's service manual.
- Switch Testing: Measure the continuity of switches, such as the ignition switch or brake light switch, to ensure they are functioning correctly.

2. Troubleshooting Electrical Circuits:

Headlights and Tail Lights:

- Voltage Measurement: Measure the voltage at the headlight and tail light sockets to ensure proper operation.

Compare the readings with the vehicle's specifications.
- Continuity Testing: Check the wiring harness for continuity to identify any breaks or shorts that could affect light operation.

Power Windows and Locks:

- Current Measurement: Measure the current draw of power window motors and door locks to determine if they are operating within the expected range. Excessive current draw may indicate a problem with the motor or mechanism.

Ignition System:

- Coil Testing: Measure the resistance of ignition coils to ensure they are within the manufacturer's specifications. An incorrect resistance reading can indicate a faulty coil.

3. Preventative Maintenance:

- Regular Checks: Use the multimeter to regularly check the condition of critical electrical components, such as battery voltage, charging system performance, and fuse integrity.
- Early Detection: Early detection of electrical issues can prevent more significant problems and ensure the reliability of your vehicle.

Home Electrical Projects

1. Electrical Outlets and Wiring:

Voltage Testing:

- Outlet Voltage: Measure the voltage at electrical outlets to ensure they are providing the correct voltage. Standard outlets should read around 120 volts (in the US) or 230 volts (in many other countries).

- Circuit Voltage: Measure the voltage at the circuit breaker panel to ensure proper distribution and detect any potential issues with circuit breakers.

Continuity and Resistance Testing:

- Wiring Continuity: Check the continuity of wiring in outlets, switches, and junction boxes to detect any breaks or shorts.
- Resistance Measurement: Measure the resistance of wiring and connections to identify high-resistance points that may cause overheating or voltage drops.

2. Circuit Testing:

Breaker Panel:

- Breaker Functionality: Test circuit breakers by measuring the voltage at the output terminals. If a breaker is tripping frequently, it may be faulty or overloaded.

Light Fixtures and Switches:

- Switch Testing: Measure continuity through light switches to ensure they are operating correctly. Verify that the switch is opening and closing the circuit as expected.
- Fixture Voltage: Check the voltage at light fixtures to ensure they are receiving power and functioning correctly.

3. Appliance and Equipment Testing:

Appliance Voltage and Current:

- Voltage Measurement: Measure the voltage supplied to appliances and equipment to ensure they are receiving the correct voltage.
- Current Measurement: Check the current draw of appliances to ensure they are operating within their rated capacity. Excessive current draw can indicate a problem with the appliance.

Grounding:

- Ground Connection: Test for proper grounding of electrical outlets and equipment. A reliable ground connection is crucial for safety and proper operation.

4. Safety and Compliance:

- Verify Ground Fault Protection: Use the multimeter to test ground fault circuit interrupters (GFCIs) and ensure they are functioning correctly to protect against electrical shocks.
- Check Electrical Codes: Ensure that all wiring and electrical installations comply with local electrical codes and standards.

Electronics Repair

1. Component Testing:

Resistors:

- Resistance Measurement: Measure the resistance of resistors to verify their values and ensure they are within tolerance. Replace any resistors that are out of specification.

Capacitors:

- Capacitance Measurement: Measure the capacitance of capacitors to check their health. Replace any capacitors that are significantly out of their rated value or showing signs of damage.

Diodes and Transistors:

- Diode Testing: Use the diode test function to check for proper forward voltage drop and reverse bias behaviour. Replace any faulty diodes.
- Transistor Testing: Measure the base-emitter and base-collector junctions

of transistors to ensure they are functioning correctly.

2. Signal Testing:

AC and DC Signals:

- Voltage Measurement: Measure AC and DC signals at various points in a circuit to verify their levels and identify deviations from expected values.

Frequency and Duty Cycle:

- Frequency Measurement: Measure the frequency of oscillators and clock signals to ensure they are operating correctly.
- Duty Cycle Measurement: Check the duty cycle of pulse-width modulation (PWM) signals to ensure they are within specification.

3. Circuit Troubleshooting:

Power Supply:

- Voltage Testing: Measure the output voltages of power supplies to ensure they are stable and within specified ranges.
- Current Draw: Measure the current draw of circuits to detect any abnormal consumption that could indicate a fault.

PCB Inspection:

- Continuity Testing: Check for continuity across PCB traces and connections to detect any broken or shorted paths.
- Component Verification: Verify the functionality of components on a PCB by measuring their electrical parameters and comparing them to expected values.

4. Preventative Maintenance:

Regular Testing: Use the multimeter to perform regular checks on critical components and circuits to identify potential issues before they become serious problems.

Component Replacement: Replace any components that show signs of wear or damage to maintain the reliability and functionality of electronic devices.

Maintenance and Calibration

Maintaining and calibrating a multimeter is crucial for ensuring its accuracy, longevity, and reliable performance. Proper care and regular calibration help prevent measurement errors and ensure that the multimeter provides accurate readings.

Caring for Your Multimeter

1. Regular Cleaning:

Exterior Cleaning:

- Dust and Debris: Use a soft, dry cloth to clean the exterior of the multimeter. Avoid using harsh chemicals or solvents that could damage the plastic or electronics.
- Screen Maintenance: Clean the display screen with a soft, lint-free cloth. For

stubborn smudges, lightly dampen the cloth with water or a mild screen cleaner.

Probe and Lead Care:

- Inspection: Regularly inspect the probes and test leads for signs of wear or damage. Replace any probes with frayed or cracked insulation.
- Cleaning Contacts: Use a clean, dry cloth or a soft brush to clean the metal contacts of the probes. If necessary, use a contact cleaner suitable for electronics.

2. Battery Maintenance:

Battery Check:

- Regular Replacement: Check the battery level periodically and replace it as needed. Low battery levels can cause inaccurate readings or malfunction.
- Battery Type: Use the correct battery type as specified by the multimeter's

manufacturer. Always replace batteries with new ones of the same type and brand.

Storage:

- Battery Removal: If the multimeter will not be used for an extended period, remove the battery to prevent leakage and corrosion.

3. Proper Storage:

- Protective Case: Store the multimeter in a protective case to shield it from physical damage, dust, and moisture.
- Avoid Extreme Conditions: Keep the multimeter away from extreme temperatures, high humidity, and direct sunlight to prevent damage.

4. Handling Precautions:

- Avoid Physical Shock: Handle the multimeter with care to avoid dropping or

subjecting it to physical shock, which can damage internal components.
- Correct Usage: Always follow the manufacturer's instructions for using the multimeter to avoid misuse that could lead to damage.

5. Regular Inspection:

- Functional Checks: Periodically perform basic functional checks to ensure that the multimeter is operating correctly. Verify that all functions work as expected and that readings are within the expected range.
- Update Firmware: If applicable, update the multimeter's firmware to the latest version provided by the manufacturer to ensure optimal performance and accuracy.

Calibration Procedures

Calibration is the process of adjusting a multimeter to ensure its readings are accurate. Proper calibration involves comparing the multimeter's readings to known standards and making adjustments as necessary. Here's a detailed guide on how to calibrate your multimeter:

1. Calibration Preparation:

- Calibration Standards: Obtain accurate calibration standards, such as precision voltage, current, and resistance sources. These should be traceable to national or international standards.
- Environmental Conditions: Perform calibration in a controlled environment with stable temperature and humidity conditions, as these factors can affect measurements.

2. Voltage Calibration:

Set Up:

- Connect the multimeter to a stable voltage source that is known for its accuracy.

- Set the multimeter to the appropriate DC or AC voltage measurement mode.

Calibration Procedure:

- **Compare Readings:** Measure the voltage with the multimeter and compare it to the known value of the voltage source.
- **Adjust Calibration:** If the multimeter's reading deviates from the known value, adjust the calibration settings according to the manufacturer's instructions. Some multimeters have internal calibration adjustments accessible through service menus or trim pots.
- **Verification:** After adjustment, verify the calibration by measuring different voltage levels to ensure the multimeter reads accurately across the range.

3. Current Calibration:

Set Up:

- Connect the multimeter to a precision current source.
- Set the multimeter to the appropriate DC or AC current measurement mode.

Calibration Procedure:

- Compare Readings: Measure the current with the multimeter and compare it to the known value of the current source.
- Adjust Calibration: Adjust the calibration settings if there is a discrepancy. Follow the manufacturer's calibration instructions to make necessary adjustments.
- Verification: Verify the calibration by measuring different current levels to ensure accuracy across the range.

4. Resistance Calibration:

Set Up:

- Connect the multimeter to a precision resistance standard or known resistor.

- Set the multimeter to the resistance measurement mode.

Calibration Procedure:

- Compare Readings: Measure the resistance with the multimeter and compare it to the known value of the resistor or resistance standard.
- Adjust Calibration: Make adjustments to the calibration settings if needed, following the manufacturer's instructions.
- Verification: Check the multimeter's accuracy by measuring resistances of different known values to ensure it performs accurately across the range.

5. Frequency and Duty Cycle Calibration:

Set Up:

- Connect the multimeter to a known frequency or duty cycle signal generator.

- Set the multimeter to the frequency or duty cycle measurement mode.

Calibration Procedure:

- **Compare Readings:** Measure the frequency or duty cycle with the multimeter and compare it to the known value from the signal generator.
- **Adjust Calibration:** Adjust the multimeter settings if there is a deviation, according to the manufacturer's instructions.
- **Verification:** Verify the calibration by measuring different frequencies or duty cycles to ensure accuracy across the range.

6. Temperature Calibration:

Set Up:

- Connect the multimeter to a temperature sensor or thermocouple with a known temperature reference.

- Set the multimeter to the temperature measurement mode.

Calibration Procedure:

- Compare Readings: Measure the temperature with the multimeter and compare it to the known temperature of the reference.
- Adjust Calibration: Adjust the calibration settings if necessary, following the manufacturer's calibration instructions.
- Verification: Check the multimeter's accuracy by measuring temperatures at different points to ensure consistent performance.

7. Documentation and Records:

- Calibration Records: Maintain records of all calibration procedures, including the standards used, adjustments made, and the date of calibration.

- Certificate of Calibration: For formal calibration, obtain a certificate from a calibration lab that verifies the accuracy of the multimeter.

8. Professional Calibration Services:

- External Calibration: For highly accurate or critical applications, consider sending the multimeter to a professional calibration service. These services have specialised equipment and expertise to ensure precise calibration.

www.ingramcontent.com/pod-product-compliance
Lightning Source LLC
Chambersburg PA
CBHW071834210526
45479CB00001B/132